THE

ART FACE OF THE UNIVERSE

A sneak peek of the abstraction side of the ocean above us.

NAVANEET JAGGANNATHAN

For permissions or inquiries, please contact:

Email: npj2309@gmail.com

Website: https://navaneetj239.web.app/Contact

This is a work of philosophical and speculative non-fiction. Portions of the editorial development were refined with the assistance of AI-based tools, under the creative direction and authorship of Navaneet Jaggannathan.

Cover Design: Navaneet Jaggannathan

Interior Design & Layout: Navaneet Jaggannathan

First Edition – 2025

ISBN: 978-93-343-4455-4

"This book is dedicated to everyone who lives under the ocean of the cosmos..."

PREFACE

Our universe is like a song—and we are the ones singing its verses. Yet, each of us sings with our own tone and tune, don't we? That means each verse carries countless rhythms, each uniquely our own.

We perceive the universe through light, sound, and feeling. The light allows us to witness both brightness and the shadow. The sound connects us to distant realms. And feelings—oh dear! —those are perhaps the most mysterious of all. Or can we? Can we ever truly grasp that depth? Think about it.

Since the days of the earliest humans, wonder about the great ocean above us has flourished. With every step forward in our understanding, a new branch blossoms—each one part of this universal symphony.

Physicists have long sought to define the universe with their brilliant, melodic theories—mathematical, elegant, and precise. But few explore the artistic side—a quiet, often overlooked perspective. One that questions the known and paints the cosmos with a new kind of light. Why not follow that path? Think about it.

As Lao Tzu once said, *"The journey of a thousand miles begins with a single step."* Let us now take that first step—together—into a journey through time and existence. Let us sing the poem of our universe with artistic rhythm and question it with our own voice.

Happy reading!
Navaneet Jaggannathan

DISCLAIMER

The Art Face of the Universe is a work of creative philosophy, artistic speculation, and personal reflection. It is not intended to serve as a scientific textbook, religious scripture, or academic thesis.

The thoughts presented throughout this book—including original proposals such as the **Brahmanic Recurrence Hypothesis** and concepts like the **Aura Theory**, and the metaphorical exploration of reality, time, death, and divinity—are hypothetical and exploratory in nature.

While the book is inspired by ideas from modern physics, quantum theories, Hindu Vedic cosmology, and collective psychological thought, it is not meant to represent any official scientific consensus or religious authority. The scientific references, such as those to the Observer Effect, quantum mechanics, relativity, or string theory, are interpreted artistically and philosophically, not empirically.

The views and interpretations shared are solely those of the author, **Navaneet Jaggannathan**, expressed in the spirit of curiosity and wonder. They do not claim to offer verified proof, nor should they be taken as guidance for personal, psychological, spiritual, or existential decisions. This book is not intended to provide medical, mental health, or religious advice of any kind.

Deep respect is extended toward all cultures, traditions, and systems of belief—especially the Vedic and Hindu philosophies referenced throughout the book. These cultural ideas are acknowledged as sacred inheritances from centuries of spiritual thought and are not claimed as personal inventions.

Readers are encouraged to approach this book with an open, questioning mind, and to draw their own meanings from its content. This work is meant to inspire **reflection, dialogue, and imagination**, not to dictate universal truth.

Additionally, while the content and voice are entirely authored by Navaneet Jaggannathan, aspects of the manuscript—such as structural refinement, grammar polishing, tone balance, and scientific consistency—have been supported using **AI tools** (including OpenAI's ChatGPT). These tools served only as editorial and developmental aids; the core insights, worldview, and narrative remain the sole intellectual property of the author.

Finally, every effort has been made to present ideas with care and respect. If any portion of this book unintentionally misrepresents, offends, or overlooks cultural sensitivities or scientific nuances, the author humbly acknowledges that such gaps are a natural part of human imperfection—and invites open dialogue.

This book is an artistic thought experiment. Not an answer sheet. Not a manual. Not a doctrine.

Thank you for reading it as such.
Navaneet Jaggannathan

ROAD MAP OF OUR JOURNEY

"

Some people see stars shine but some people see stars extinguish!

Abstraction and Logic are the two opposite poles of the same magnet.

"

… Every second, approximately,

A hundred twenty Black holes are born, sixty million planets are formed, twelve hundred stars explode, and sixty thousand stars are born. Our Milky Way Galaxy travels a thousand kilometres and grows by five hundred meters, and while we gaze at the sky, our Universe has expanded by two million kilometres! Think about it.

Source: https://youtu.be/psJbhQuKyIc?si=4ZacPeJoHh9ZJtgl

#THE CREATION AND THE DESTRUCTION

The question of how the universe began—and how it might end—has fascinated minds across time. From the mystical hymns of ancient sages to the mathematical equations of modern physicists, the quest to understand the universe's origin is a thread that runs through our collective consciousness. But with all that effort, has anyone truly deciphered the first verse of this cosmic symphony? Or the final note?

Many physicists have tried to compose their own versions of this grand story. Theories have emerged like melodies—some harmonious, some discordant. Among the most discussed are the Big Bang–Big Crunch, the Big

Bounce. It tries to describe the universe's birth and death through the lens of physics. But is that the only lens?

After all, can we not view this cosmic play as something more than equations and measurements?

Can the universe not also be seen as art—rhythm, contrast, repetition, transformation? Isn't the act of creation and destruction itself a kind of celestial choreography?

Let us begin by exploring what the current science says—and then we shall move toward a new perspective, one rooted in ancient wisdom, and reimagined through a modern lens: **the Brahmanic Cycle**.

The Big Bounce: A Dance of Fire and Collapse

Among the many verses written by modern science, the Big Bounce stands out as one of the most poetic. It is a theory that says the universe is not a one-time phenomenon, but rather a cycle—a great heartbeat of cosmic time.

In the Big Bounce model, the universe begins with a fiery explosion—the Big Bang—expanding from a point of infinite density and temperature.

As time passes, expansion slows, and the gravitational pull of all matter causes the universe to contract once more. Eventually, everything collapses back into a dense singularity—the Big Crunch.

But this is not the end.

From that singularity, a new universe is born. The process begins again—Bang, expand, contract, Crunch...

and bounce. This rhythm continues infinitely, like a song that never ends, playing out on the largest possible scale.

The Big Bounce challenges the standard inflationary model, which posits a one-time Big Bang followed by continued expansion.

It tries to resolve some lingering problems in cosmology, such as what came "before" the Big Bang, and what happens to information in a collapsing universe.

But even this elegant theory, for all its cosmic poise, leaves a question: *what causes the bounce?* And what lies between and beyond the cycles?

THE BRAHMANIC CYCLE: A VEDIC ECHO IN MODERN THOUGHT

Now, let us shift the lens. Imagine a universe that doesn't simply begin and end—but **loops eternally**, not as repetition, but as **reinvention**. Imagine time not as a straight line, but as a spiral, or an ocean of random tides—each turn carrying echoes of the last, yet unfolding in its own unique form. In this vision, destruction is never final—it is **transformation**. A new form, a new rhythm, a new dance.

This is the heart of the **Brahmanic Cycle**, a hypothesis I wish to propose—***'The Brahmanic Recurrence Hypothesis'***, a theory inspired by the ancient echoes of **Hindu Vedic cosmology**, and reinterpreted through the evolving lens of modern science.

In the Vedas and Puranas, the universe is portrayed as existing in vast cycles of **creation, preservation, and destruction**. These are embodied in the divine triad: *Brahma* the Creator, *Vishnu* the Preserver, and *Shiva* the Destroyer, the celestial and symbolic representations of the ultimate reality, *Brahman*. The cosmos lives through cycles called **Kalpas**, each spanning billions of years. At the end of each Kalpa, the universe dissolves into a latent, unmanifested state—only to be reborn again from that stillness. Such a magnificent concept has been personified in the old Vedic texts and puranas.

This ancient cosmology doesn't imagine a linear end or beginning, but a **cosmic rhythm**—a pulsing of the universe itself. Not repetition for repetition's sake, but **renewal** born from necessity. A phoenix rising not from ashes, but from inevitability.

The **Brahmanic Cycle** builds upon this spiritual heritage and reframes it in the language of **cosmology**, **entropy**, and **quantum possibility**.

FOUNDATIONS OF THE BRAHMANIC CYCLE

The Brahmanic Cycle begins with a core premise:

'The universe has no true beginning and no absolute end—only phases of transformation.'

It holds that all **matter and energy are conserved**, but more importantly, so is **cosmic potential**. When our universe reaches its long-projected "end," whether through cold entropy (the Big Freeze) or an accelerated expansion (the Big Rip), the components of the cosmos don't disappear. They fade, stretch, and scatter into what I call a **fall-state**—a realm of near-zero temperature, low energy, and maximum entropy. It is a universe stripped of light, motion, and structure.

But here lies the paradox—and the possibility.

In this dark silence, the **vacuum is not empty**. Even in the most lifeless void, **quantum fluctuations**—tiny, random disturbances in spacetime—continue. They are an inescapable feature of the quantum world, arising from the Heisenberg uncertainty principle. These fluctuations, invisible and unpredictable, are the **everlasting disturbance** that underpins the next chapter of existence.

Over unimaginably vast spans of time—trillions upon trillions of years—these quantum ripples can accumulate. They can, under the right conditions, **spark the spontaneous birth of a new universe**. This isn't mere speculation; it mirrors ideas from frontier cosmology, including vacuum decay, baby universes, and quantum tunnelling.

In the Brahmanic Cycle, **a new universe doesn't emerge from nothing**—it is seeded from the instability left behind by the previous one. The chaos of the old becomes the canvas of the new.

It's not reincarnation in a metaphysical sense—but it carries its echo. It's not exactly physics as we know it—but it grows from its soil.

This vision is not the eternal silence of the Big Freeze, nor the final scream of the Big Rip. It is **continuity**. Not of form, but of essence.

The **future of one universe becomes the past of the next**—woven together by time, by tension, and by timeless disturbance.

Just as a forest regrows from fallen leaves and roots long buried, so too does the universe **blossom again** from its own ruins.

THE BRAHMANIC CYCLE VS. EXISTING MODELS

At a glance, the **Brahmanic Cycle** might appear similar to the **Big Bounce**, but their philosophies—and mechanics—differ.

The **Big Bounce** relies on **gravitational contraction**. After expanding for eons, the universe reverses course, collapses into a singularity, and then rebounds into a new expansion. It's a rhythmic heartbeat of geometry—bang, crunch, bounce, repeat.

The **Brahmanic Cycle**, however, proposes something subtler—and perhaps more enduring.

- It does **not require contraction**.
- It works even in eternally **expanding universes**.

- It does not demand an external force to collapse the cosmos, but instead relies on the **internal instability of space-time**.

The transformation is **not geometric, but energetic and informational**—carried through entropy, fluctuation, and conservation. Even when matter has decayed, when stars have died, and black holes have evaporated, the **quantum fabric remains restless**. That restlessness is the echo of the previous cycle and the birth cry of the next.

Crucially, the Brahmanic Cycle honours the **law of conservation**—not only of mass and energy, but of **possibility**. It suggests that no universe dies in vain; every cycle leaves behind just enough disturbance, imbalance, and latent energy to reimagine itself anew.

A UNIVERSE OF VERSES

When viewed through this lens, the universe is no longer a singular story. It is a poem with infinite stanzas, each one rhyming with the last, but never repeating word for word. Each cycle is a verse; each transformation, a rhythm.

From the fires of one ending comes the spark of another beginning. From silence, song. From ash, light.

So, what if our future is already written—in a past we cannot remember? And what if our destruction is simply a whisper of rebirth?

Right? Think about it.

#A QUESTION OF REALITY AND FANTASY

Have you ever truly paused and asked yourself—*is this real?* Not just in a metaphorical or dreamy sense, but in the most basic, unsettling way: *Is the life I'm living... real?* Are the sensations I feel, the thoughts I think, the memories I hold—are they actual events anchored in some objective truth? Or is there a possibility that we're living inside a simulation, a constructed illusion so perfect, we mistake it for reality?

If you haven't asked that question yet, don't worry—someone else has. In fact, many people have. Scientists, philosophers, mystics, even Silicon Valley billionaires.

But before diving into theories of simulations and artificial matrices, let's rewind to something much more personal. Why is it that **we**, beings inside this universe, even have the capacity to imagine something *outside* of it?

That's the real mystery. Right? **Think about it.**

PERCEPTION AND REALITY

Let's start simple. Take a coloured photograph—let's say a picture of a sunset. You show it to a human, and you show the same picture to a dog.

Will they see the same thing?

The human might admire the rich oranges, pinks, and purples bleeding into one another. They might feel something—nostalgia, awe, warmth. The dog, however, won't see red or green. Dogs don't process colour the way we do. Their vision is dichromatic—they mainly perceive shades of blue and yellow. So, while the human sees a warm cascade of hues, the dog sees a duller, two-toned landscape.

So, who's right? The answer is: both.

The sunset didn't change. What changed was the **perception**. The same photons hit both eyes, but the interpretation—the brain's response—was entirely different. Colour, then, isn't a universal truth. It's a **subjective creation**. It exists not in the object itself, but in the interplay between the object, light, and the observer's biology.

So, does this mean reality is just... how we see it?

Or that what we call "reality" is merely a **consistent fantasy** shared among similar observers?

Once again—**think about it.**

THE DANCE OF REALITY AND FANTASY

Let's try to define two things clearly before they blend too much:

- **Reality** is that which continues to exist whether or not we believe in it. It's stable, objective (as much as we can agree on), and verifiable. It has structure.
- **Fantasy** is created in the mind. It's imagination. Subjective. Fluid. It doesn't have to follow the laws of physics—or even logic.

But here's the twist: they aren't enemies. They don't live in separate realms. They exist *within each other.*

Let's break that down.

A **fantasy in reality** is when your real-world experience seeds an imaginative offshoot. You see a blank wall, and in your mind, you envision a portal. You know it's not there, but your imagination has created a fantasy *inside* reality.

A **reality in fantasy** is trickier. It's when your imagination becomes so detailed, so compelling, that it begins to influence your behaviour, your emotions—maybe even your worldview. It's when belief takes root in imagined soil. This is how religions are born, or at least as per the atheist and extreme philosophers! This is how great inventions start—as ideas, sometimes wild and unreasonable, that slowly manifest into tangible things.

Now let's anchor this with two scenarios.

SCENARIO 1 – THE IMAGINED VICTORY

Picture a warrior on the battlefield. Around him: chaos, noise, uncertainty. He closes his eyes for a moment, breathes in deeply, and tells himself, *"I have already won."*

At first glance, this may seem like denial. A fantasy. But something remarkable happens. That internal declaration—imaginary as it is—begins a cascade inside him.

His brain responds. Confidence rises. Adrenaline flows. Cortisol drops. His posture changes. His vision sharpens. His focus becomes narrow, precise. Decisions that would have been panicked about now come with clarity.

His belief—though not based on current truth—has **shaped** his physiology. His **fantasy has altered his internal reality**.

And perhaps, in a battle of moments and margins, that shift is enough to survive.

He didn't *win the war* in that moment. But his imagination made him *capable of winning it.*

That's not just psychological—it's biological. It's a **fantasy inside reality** that changes the outcome of reality itself.

SCENARIO 2 – THE PAINTED POSSIBILITY

Now, picture a blank canvas. You mix blue paint with yellow, and green appears.

Simple, right?

But pause for a second: that "green" was nowhere to be found in either original colour. It emerged from the act of blending. And you expected it—because you've seen it before or someone has said you before. But your expectation was rooted in **experience**, not certainty.

Here, your *fantasy* (expectation of green) was confirmed by *reality* (the laws of pigment interaction). So, it's not imagination *against* truth—its imagination informed by it.

TIME: THE FINAL METAPHOR

Let's bring in the concept of time.

Newton imagined time as a straight, rigid arrow—flowing equally for all, moving in one direction, never changing speed or shape.

Then came Einstein, and he shattered that certainty. Time, he said, is more like a river. It bends. It speeds up or slows down depending on gravity and motion.

In fact, two people experiencing the same moment might actually be moving through time at different rates. Time is not one universal truth—it is relative.

Now that's not just science. That's poetry.

Because what Einstein did wasn't just redefine equations—he redefined **reality** itself.

What we once thought was fixed became fluid. What we called the objective truth turned out to be relative to the observer.

Just like with colour. Just like with experience.

WHAT IS THE REAL DIFFERENCE?

Fantasy isn't the opposite of reality. It's the beginning of it.

We imagine before we create. We dream before we build. The spaceship is a fantasy—until someone makes it. The equation is imagination—until it describes nature.

Our entire progress as a species is a trail of fantasies that *became* reality.

And sometimes, we carry realities inside fantasies, too. We dream of things that never were but draw from everything we know—memories, emotions, hopes. So, ask yourself:

Are you living in a reality that contains fantasy? Or are you living in a fantasy that shapes your reality?

What if the answer… is both? Think about it.

#THE MIND AND TIME

The mysteries of the mind and the nature of time—two of the most elusive and enduring puzzles known to us.

We've spent centuries trying to measure them, define them, and understand them.

And yet, they remain just beyond full comprehension—like trying to capture fog in our hands.

So, let us now step into another layer of the universe's enigma, where thought meets chronology. Where inner perception dances with cosmic order.

Where **mind and time** are not opposites—but mirrors.

TIME: THE MAP OF EXPERIENCE

When we think of time, most of us recall Einstein's iconic words:

"Time is relative."

A simple phrase, yet it unravels the foundations of how we perceive existence. Einstein showed us that time does not flow the same for everyone. It can stretch and bend depending on speed, gravity, and position.

This is known as **time dilation**—where time moves slower for someone traveling close to the speed of light, compared to someone standing still. For astronauts near a massive gravitational field, minutes may pass while years go by on Earth.

But let's zoom in from space to something more personal.

Time isn't just a number on a clock—it's an **experience**. Our brains don't perceive time the way a machine counts it.

Instead, time flows faster when we're immersed in joy, and slower when we're waiting, anxious, or bored.

That's not a poetic metaphor. That's **neurology**.

Your brain measures time based on attention, emotion, and memory. The more sensory input and emotional variation, the **shorter** a period *feels*—even if it's objectively longer.

So, in a sense, **your experience of time is shaped not just by the laws of physics—but by the lens of your mind.**

THE MIND: A GATEWAY OF CONNECTION

What is the mind?

Some call it a machine made of neurons. Others see it as a window to consciousness, memory, intuition, and imagination.

But here, let us see it as a **bridge**—a medium that connects perception to experience. A translator between **what is** and **what we understand**.

When we feel love or affection for someone, it's not our hands or words that create that bond—it's the mind that **recognizes** the emotion and **builds** the connection.

Affection, then, is more than a feeling. It's a **mental relationship**—a channel between you and another, shaped by your mind's perception of time, memory, and meaning.

Even speculative ideas like **telepathy** or **mind-matter interaction**—though unproven—stem from this same yearning: to connect, to influence, to relate.

Whether or not those are real phenomena, the symbolism still holds:

'The mind is a tool for relating. Time is a medium through which those relationships unfold.'

HOW THE MIND AND TIME INTERTWINE?

Now here's where it gets interesting.

What happens when we combine the mind and time? We don't just get memory—we get **meaning**. We get the sense of a personal story unfolding in a unique rhythm. A subjective timeline.

Let's take an example.

Imagine two people in the same room—both waiting for the same event. One is excited. The other is anxious.

To the first, the time flies. To the second, it drags endlessly.

Same clock. Different minds. So which one is real?

In this moment, **time is no longer just external. It's internalized. Personalized:**

"If you're idle, time feels slow. If you're doing something you love, or have expectations, time feels fast."

It sounds simple—but this speaks to a **fundamental interaction** between brain chemistry and perception.

When you're bored, your brain registers fewer sensory inputs. Your awareness lingers on time itself, stretching each moment.

But when you're deeply engaged—reading, playing music, laughing with a friend—your attention is diverted. The brain enters a state of **flow**. Neurotransmitters like **dopamine** and **serotonin** rise. Time, as a feeling, begins to vanish.

So, we must ask:

Is time truly slow or fast—or is it the **mind** that alters its texture?

A THOUGHTFUL METAPHOR: MOTHER AND SON

"The past of a mother giving birth to her son create the future of his life."

This is a poetic and powerful truth.

A mother's experiences shape not just her own timeline—but directly influence the child's. Her memories, choices, and even her biological state during pregnancy—shape his world before he's even aware of time.

So, the timeline of one person becomes the **foundation** for the timeline of another.

Time here is not linear. It's **relational**. It flows **between people**, not just within them.

FINAL REFLECTION: TIME, MIND, AND MEANING

Time gives events a place. The mind gives them a purpose.

Both are invisible, immeasurable in full, and yet undeniably real in our experience.

So, when we say "time is relative," we must ask: **Relative to what?** To gravity? To motion? Yes. But also—**to the mind itself**.

Because a minute can feel eternal. A year can vanish in a blink.

And so, the truth is this: ***The mind doesn't just experience time. It creates it.***

Think about it.

#A RANDOM OR A SCRIPT?

Aha! We all love watching movies—and there's something magical about peeling back the curtain to discover the director's vision, the carefully crafted plot, the blueprint behind every scene.

But have you ever paused and wondered: **When and how did the director come up with that scene's plot in the first place?** Was it fully scripted from day one or born in a spark of improvisation?

Think about it.

IS LIFE A MOVIE—OR AN AD LIB?

Every film we watch follows a chain of scenes—connected by cause and effect, intentional design, or the chaos of reality itself. We see plans play out. Plot twists. Moments of spontaneity. The filmmaker either directs us through a boilerplate script or leaves room for actors' instincts and unplanned brilliance.

This makes us ask: **Is our universe like that—a carefully written screenplay—or is it more like an improvised sketch, crafted moment to moment?** Consider this:

- If it's scripted, then who or what made the blueprint?
- If not, then every moment we experience—every breeze, every meeting—becomes part of a cosmic improvisation.

Are we stirred by fate, randomness, or a mix of both?

Think about it.

BLUEPRINTS IN PHYSICS: THE SCRIPTED UNIVERSE

Modern science leans heavily on a **scripted universe** model. From Newton to Einstein, physics suggests the cosmos follows set laws.

But we live in a realm where **laws are constants**, codifying cause and effect. Gravity, electromagnetism, and subatomic forces—all play their role. Space and time obey their own syntax.

But here's the catch,

The real catch that may sound vague, but always questions me,

If everything is scripted, why can't we predict our distant future? Why can't we forecast thc weather a year from now? How could seemingly random events—like volcanic eruptions or falling stars—still surprise us?

CHAOS THEORY: SPACE SCRIPTS, BUT TIME'S A MISTRANSLATOR

Even with unbreakable laws, our **predictive power is limited** because:

In **complex systems**, tiny differences grow exponentially. A butterfly flaps its wings, and a hurricane shifts course weeks later.

This is **chaos theory**, where **order and disorder coexist**, and where predictability crumbles under complexity.

We can map the structure—the "blueprint" of existence—but we can't foresee every twist. Space has boundaries we understand, but time? We write the script one word at a time—all while wondering where the next line will come from.

QUANTUM PHYSICS: THE UNIVERSE ROLLS THE DICE

Now, zoom to the smallest scale of reality—atoms, electrons, quantum fields. Here, the universe doesn't play by Newtonian rules at all:

- Events are **probabilistic**—not predetermined.
- **Quantum indeterminacy** means particles don't have exact positions or speeds until measured.
- Einstein famously resisted this: *"God does not play dice."* But the universe seems to disagree—we see experiments that confirm randomness is woven into its fabric.

So, our cosmic film may follow physics' structure—**but with improvisation cues built in**.

SCRIPTED RANDOMNESS: A HYBRID STORY

What if the universe isn't strictly scripted or completely chaotic—but a **creative interplay** of both?

Picture it like this:

The **laws of physics** provide an overarching framework—a universe-wide screenplay, if you will.

But within that script, there's **space for improvisation**—quantum randomness, chaotic branching, emergent complexity.

So, the overall narrative unfolds according to rules. But the details—who we meet, when, what we choose—they are not fully known until they happen. They are the universe's **everyday improvisations**, waiting to emerge.

Think of it this way: Maybe the universe is like a long-running musical jam session. The chord progression is defined, but the melody evolves in real time.

COMPRESSION, ENCRYPTION, AND EMERGENCE

"Predefined randomness"—like encrypted data, where you see chaos, but underneath it's following a signature, a pattern.

This reminds me of how:

- A **random mathematical seed** can generate a coherent image.
- Or how **in fractals**, a simple iterative rule produces intricate forms of surprising beauty.

Does this mean there's a script inside every apparent random moment? Possibly. It suggests that **randomness may not be meaningless**, but could still follow hidden structure or direction.

THE ARTISTIC PERSPECTIVE

Returning to our filmmaker analogy: Movies can be made in two styles:

1. **Fully scripted**: Every line, camera angle, background cue—planned.
2. **Mostly improvised**: Actors and crew shape scenes on the fly, responding to emotion, environment, or mood.

Yet both produce something cohesive.

What if the **universe** works the same way?

At its foundational level, there's a script—physical laws. As the universe unfolds, improvisation occurs—quantum events, complexity, choice.

And maybe **we**, as conscious beings, are part of that improvisation. Our decisions—truly new moments—are **co-writing** the cosmic screenplay. **Think about it.**

Where the Script Ends—and We Begin

Remember the last time you were fully in the moment—maybe playing music, painting, or having a heartfelt conversation. In that instant, **time expanded or collapsed**, and life didn't feel bound by a script. Nothing felt random either. It was **co-created**, by you and the moment. The outline was there—but the melody was never fully written before you began.

This is permission for your own creative authority. You're part improvisor, part observer, part composer.

To tie all this together:

- **Blueprint exists in physical law**—space, time, gravity, physics.
- **Complexity and chaos** introduce surprise and spontaneity.
- **Quantum mechanics** affirms randomness at core.
- And **conscious processes**—choices, creativity, awareness—end up playing a role in unfolding events.

But none of it can be known ahead of time.

So, **we exist within a universe that is both scripted and free**. We're not completely pawns. Nor are we uncontrollable anomalies.

We are **co-authors**—fragments of creativity writing literally nothing less than our existence.

FINAL REFLECTION: WHAT DOES IT MEAN TO LIVE FREELY?

What if knowing there's a cosmic "plot outline" gives you security—but knowing there's openness in the details gives you freedom? What if living is not just following the script, but **choosing your rhythm within it**?

The universe grants space for spontaneity. It rewards improvisation. And perhaps, the most exquisite truth is that **every moment you truly act—not just react—is a plot twist in the universal story.**

Is your life random, scripted, or both? Think about it.

#A DREAM AND THE MIRROR

Have you ever woken from a vivid dream and found yourself questioning: *Which world was the dream—and which is reality?*

The images, emotions, and voices felt so real... but they vanished when you opened your eyes. What if **all** of reality is a dream? Or perhaps a mirror reflecting deeper truths?

Think about it.

THE QUANTUM INTERSECTION: WHERE OBSERVATION SHAPES FORM

In the strange world of the very small, physicists have discovered something extraordinary. At the heart of quantum mechanics lies a simple yet profound puzzle: **the observer shapes reality**. Walk into a dark theatre with a double-slit apparatus and cast a glance, and suddenly, a particle behaves like a particle. Leave it alone, and it becomes a wave. Watch it pass through a slit, and the wave collapses into a pinpoint dot. Observe its path—and the universe seems to obey.

This isn't magical thinking. It's a scientific observation backed by countless experiments. The double-slit acts like a filter: without observation, the universe exists in shimmering waves of possibility. With it, it collapses into the outcome we witness.

What this tells us is profound: **reality doesn't declare itself until someone asks the question**. The universe holds countless possibilities—but it doesn't commit to any until we observe. Like a painter holding back their brushstroke until the canvas demands it.

'Until we perceive, reality remains fluid, like uncharted territory.'

And so, the line between dream and waking life blurs: if our gaze collapses the wave, then maybe reality *depends* on us. Maybe our entire existence is the universe *experiencing itself*. Perhaps, with each moment of awareness, we sculpt a facet of reality and then step away.

MAYA: THE VEIL OF PERCEPTION

This idea isn't new. Ancient traditions, particularly within Vedic philosophy, described a similar insight: **Maya**, the world as illusion. Not illusion in the sense of deception or error, but illusion as a **veil**—the world of appearances shaping our perceptions. Maya creates form, time, and space—perceptual structures that make experience coherent, but which also distort the deeper truth.

The earthly world becomes a stage with actors playing parts. We see masks, not faces beneath. We hear noises, not cosmic songs. We feel separate from everything, even as physics suggests a deep interconnection between all things.

Here, Maya meets quantum mechanics. Behind each wave-particle duality, each collapse of probability, lies a hidden unity—did you recognise it, dear reader? — that may only be made clear when we awaken from illusion.

THE MIRROR WITHIN: JUNG AND THE REFLECTED SELF

Swiss psychiatrist Carl Jung brought this cosmic mystery into the terrain of the human psyche. In his work, **the unconscious mind is a mirror**, reflecting back hidden aspects of self, potential, and shadow. We gaze into the unconscious, and it returns symbols, dreams, impulses—a hidden cosmos of the psyche.

Jung spoke of the *collective unconscious*—archetypal forces shared among humanity, like fountains that feed each individual's psyche. These forces not only shape our inner worlds, but our outer ones, too—our myths, stories, our understanding of God, cosmos, and mortality.

It's a two-way mirror. Just as we look outward to perceive the world, the world—and perhaps *the universe itself*—looks inward at us, through the lens of consciousness. The cosmic mind reflects in the human mind, and vice versa, like infinite mirrors facing each other.

And so, the dream becomes real in reflection, and the real becomes a dream in interaction.

THE PARADOX OF ILLUSION AND REALITY

Are we all just characters inside a dream, mirroring a cosmic mind?

And yet… If no one dreams, does the world cease to exist? If observation births collapse, what of silent galaxies, drifting without witness? Scientific consensus says observation doesn't require human consciousness—

it just requires **interaction**, entanglement with another system measuring it.

But the philosophical echo remains: **observation creates form**. Whether a particle is measured by a photodetector or a human, the wave function collapses. Reality takes shape. It's as if the cosmos mirrors itself wherever attention shines.

Illusion becomes creativity in the hands of the observer. And perhaps **paradox** is the canvas. If the wave holds all possibilities until observed, then unobserved reality remains as rich as dream—not less real, just not differentiated. Perhaps the unobserved is as vital as the observed, like the sky behind the clouds.

THE UNIVERSE VISUALISED: EXPERIENCING ITSELF

Have you ever looked at a snowy field at dawn and felt the quiet electricity of potential? Before footsteps, tracks, or warmth, the snow lies in pure form—silent symmetry. In that instant, the universe is *for itself.* It's unobserved by humans, yet full of possibility. It *is*. Then you step on it. Your foot collapses snow. Form emerges. Observation and interaction create reality.

And maybe that's the true art of the cosmos: **it perceives the snow even before we do**, but it waits for the moment we step—to make the message visible.

In that divine waiting, we co-create not just form, but meaning.

DREAM AND MIRROR: INDIVIDUALLY CO-WRITTEN

In your own dreams, you may travel, speak, and live without knowing they are dreams. They feel real until you wake. The waking world feels real—until you wonder if it's just a larger dream.

Consider this possibility: **everyone dreams. Everyone mirrors. Everyone collapses the wave of reality.** And yet none of these dreams is 'fake'—they are *crucial to the dance*. Whether wave or particle, dream or wake, mirror or stage, they all matter. All are phenomenally real—each in its own register of experience.

If we are the universe experiencing itself, then our dreams are the internal dimension of that experience.

They are the mirror reflecting not just ourselves but the cosmic pattern.

UNRAVELLING THE TAPESTRY OF ILLUSION

Is this all an illusion? Yes and no. Illusion in that it is shaped by observation and perception. But deeply real in that **it carries truth**, in pattern, in dance, in discovery. Maybe the truth is not that life is unreal, but that it's **more real than we can ever conceive when we hold on to being separate**. Maybe Maya and quantum mirrors remind us we aren't separate.

When you say, "I think, therefore I am," don't forget that "I" exists inside a reflection, shaped by cosmic light, hidden archetypes, and entangling particles. To wake up from the dream is not to leave it behind, but to include it consciously.

THE FINAL MIRROR: CONSCIOUSNESS OF OUR ROLE

What if the universe had no eyes—no awareness—until complex minds evolved to look back upon it?

You: a living lens. An artist's brushstroke folded inside the painting itself.

The mirrors facing mirrors… infinite reflection… is not just poetic. It is *factual*, in the sense that our consciousness, our observation, and our storytelling are entangled with reality. And this entanglement is art.

CLOSING REFLECTION

So, ask yourself, right now:

Is the scene before your eyes real, or another projection? Does the dream in your mind carry more truth than the timetable of your schedule? Is your reflection in the mirror *you*, or *something else* far vaster? Are we part of a cosmic dream, or is the dream part of us?

Think about it.

You are not simply human. You are a *universe experiencing itself.* A wave becomes a particle. A mirror sleeping to reflect. A dream that looks at its own reflection and wakes, over and over again.

And in the soft pause between wave and collapse… In that breath before you realise... You are art, woven from the fabric of cosmic consciousness.

Right? Think about it!

AFTER-PLAY OF DEATH

Death. The great mystery. The invisible gate that closes behind every life. We speak of it in whispers, in prayers, in poems.

We run from it, chase it, and at times try to understand it—but do we really know what lies beyond?

We've walked through the creation of the universe, we've danced through time, and now we arrive at the silence that follows the final breath. But is it truly silence? Or just a passage into something deeper, subtler, and more cyclic than we imagined?

Think about it.

A PARABLE OF SELF: THE EGG

One of the most thought-provoking perspectives on life after death comes not from ancient scriptures or modern science, but from a simple short story—*The Egg* by Andy Weir.

In it, a man dies and meets a divine being who explains that he is not just one person, but every person who has ever lived. Every time he caused pain or kindness to another, he did it to himself. Every birth, every death, every experience, is him—over and over—until he lives them all. Only after experiencing every soul can he transcend and be born as a god.

It's not a story about morality. It's about understanding. About compassion born from identity. It proposes that we are not individual souls, but **one soul experiencing itself from every angle**. Life becomes a loop. Death is just a portal back into the script—different face, same soul.

Now that's a thought, isn't it?

THE ANCIENT LOOP: KARMA AND REBIRTH

Long before The Egg, ancient Vedic thinkers had already been painting their own visions of this loop.

In Vedic philosophy, death is not an end—it's a **junction**. The soul, known as *Atman*, carries its karma from life to life. Good or bad, kind or cruel, every action echoes beyond the body. At the moment of death, the Atman exits the body like smoke rising from a flame and seeks its next vessel, based on the karmic residue it holds.

If one has balanced their karma—through dharma, self-awareness, and detachment—they are said to achieve **moksha**, liberation from the cycle. At that point, the Atman merges with **Brahman**, the absolute. No more births. No more illusions. No more separateness.

The Vedic vision is mathematical, moral, and mystical—all at once. It doesn't tell you who you are. It asks you who you are becoming.

But here's the catch: both The Egg and the Karma model speak of reincarnation and return. So are they two tunes in the same raga? Or fundamentally different scores?

EGG VS. KARMA: ONE SOUL OR MANY?

At first glance, these models look similar. But the deeper you go, the more you see their divergence.

The Egg speaks of a **single soul**—you are everyone. Every life is yours. The cycle continues until you've played every part in the cosmic play.

Karma, on the other hand, retains the **individual soul**—unique Atmans looping through time, each governed by their own karma.

You are you. Your journey is yours alone. Your salvation depends on the purity of your own actions.

The Egg proposes that empathy is born from identity. Karma proposes that empathy is born from accountability.

Both offer profound reflections—but what if there's another possibility?

THE AURA THEORY: A DEEPER LENS FOR THE VERY EXISTENCE

What if death is not about giving a meaning to the soul-ownership or moral judgment, but about **energetic resonance**?

Let me repropose it with a new perspective—a hypothesis that weaves through Vedic thought, absorbs echoes of modern physics, and reflects a principle of energetic flow. I call it the *'Everlasting Search for Cosmic Stability until breakdown'* or the **'ES-CSUB Hypothesis'**.

Imagine that each living being carries an aura—not just an emotional glow, but a **subtle energy field**, dynamic and reactive. This aura is made of polarities: positive frequencies shaped by compassion, truth, and balance; and negative frequencies born of fear, illusion, and desire. It's not spiritual fiction—it's the vibration of your lived experience.

As we live, we shape our aura—like painting on a canvas that stretches beyond our skin. And when we die, this aura separates from the body, floating in a delicate state of flux.

Now, what happens?

If your aura is balanced—harmonic, calm, pure—it naturally dissolves into the **field of Brahman**. Like a sound wave absorbed into silence. This is moksha. Liberation. Salvation. Union.

But if your aura carries dissonance—negative karmic charge—it becomes **unstable**, like a vibrating string seeking resolution. This energy cannot merge with Brahman until it's purified. So, it wanders—searching for a vessel, a new birth, a womb forming in resonance.

And all of this happens not across ages, but in what we may poetically call a **slit in time**—a quantum moment where birth and death intersect. In that fleeting gap, the aura finds its next shell.

This process isn't moralistic. It's **mathematical**—a dance of frequencies. The aura is not rewarded or punished. It simply flows until it stabilizes. Now think about it: Isn't that what karma always was? A system not of vengeance but **balance**?

AURA AND MAYA: THE TRAP OF THE ILLUSION

In Vedic thought, *Maya* is the grand illusion—the curtain of perception that makes the finite appear separate from the infinite. Maya traps us in the dance of names and forms, keeping the soul wandering through lifetimes.

From the Aura Theory lens, **Maya is the cause of energetic distortion**. It's Maya that draws us into attachment, violence, ignorance. The more we chase it, the more unstable our aura becomes. And so, we are reborn—not as punishment, but as a natural consequence of energetic imbalance.

Until one day, we let go. Our actions flow from stillness. Our thoughts are not reactions, but reflections. And then the aura begins to calm. The cycle slows.

And when that day comes—the aura vanishes into Brahman like dew into the sun.

Three Visions, One Raga

Let us pause and compare—not to compete, but to harmonize.

The Egg tells us: “You are all beings. You must live every life to transcend.”

Karma says: “You are one soul. Live rightly, and you will be free.”

Aura Hypothesis whispers: “You are energy. Balance yourself, and you return to source.”

One speaks of **unity through identity**. One, of **freedom through virtue**. One, of **stability through resonance**.

All are music. Different instruments. Different rhythms. But perhaps they are all playing the same cosmic song.

DEATH NOT AS AN END, BUT A TRANSITION

So, when we ask, “What happens after death?”—perhaps we shouldn’t ask *where* we go, but *how* we transform.

Are we recycled? Are we renewed? Are we remembering?

Maybe death is not a door, but a mirror. Not the falling of the curtain, but the changing of scenes. The actor remains. The costume changes.

And if your aura flickers with imbalance, it finds another stage. If it rests in stillness, it fades into the silent music of the cosmos.

Now think about it again.

What if the universe doesn’t end with your last breath? What if, in a slit of time, the story just begins again?

EMOTION’S A GEOMETRY

Imagine standing at the edge of a vast canyon—your heart races. The wind sweeps past you. The chasm seems both terrifying and exhilarating.

That moment of awe—when your pulse hammers and your breath catches—what is it? Is it simply chemicals in your brain... or is it something more intricate, something shaped like the canyon, shaped like your own inner world? Could emotions themselves have geometry? Think about it.

CURVATURE IN MIND AND MATTER

In physics, Einstein revealed that gravity is not an invisible force pulling us down. Instead, mass curves space-time, and we fall toward that curve. A planet, a star, or even an emotion—they all create a shape in their own medium.

Emotions may not curve galaxies, but they curve *us*—our attention, our posture, our thoughts.

Love might feel like gentle valleys, warm and spacious. Grief, a steep ravine of dense melancholy. Anxiety could be jagged peaks rising abruptly in your chest. Joy might expand like infinite plains.

Our internal landscapes mirror the external ones. And if space-time is shaped by stars and mass, could our emotions be shaping inner space-time?

EMOTIONAL TOPOLOGY: EVERY FEELING HAS FORM

Topology studies how surfaces and shapes behave—how a coffee mug can be reshaped into a doughnut, because each has one hole. Emotions, too, might be topological: the fold of worry, the hole of loneliness, the smooth surface of calm. We can't erase them—they have form, but we can transform them.

Consider shame: a deep crumple in your inner map, hard to flatten until you gently reshape it. Contrast that with love: a soft plane, easy to walk across. Each emotion has borders, layers, and depths.

And sometimes they overlap. You can feel sad and grateful—folds interlocking. Emotions aren't separate—they're multidimensional.

THE GEOMETRY OF EMOTIONAL EXPERIENCE

When something happens—a kind word, a loss, a memory—the shape in us changes. A smile creates an expansion, a sigh maybe a hollowing. Our minds adapt and reshape in response.

Neuroscience shows us how emotions stem from networks firing in different brain regions. But those networks are also spatial: neurons connecting, pathways strengthening. The geometry of emotion is literal: circuits and networks creating patterns that echo the shapes we feel inside.

It's not just metaphor—it's physical, embodied geometry.

GRAVITY OF EMPATHY, ORBITS OF LOVE

Some emotions pull us deeply. Empathy draws you inward toward another's pain until their sorrow begins to feel like your own. That's empathy's gravity—a bond you can't easily escape. You're caught in the curve of their emotional space.

Then there's love at a distance—like orbiting a planet. You influence each other, but maintain your own paths, your own arcs. Close, connected—but free to circle.

Friendship might be like kinetic energy—fleeting, swirling, shimmering. Passion could feel like rapid pulses of light and heat. Everything moves. Everything is shaped.

PAIN AS GEODESIC DEVIATION

In general relativity, two objects moving along curved space-time can start together but drift apart or converge depending on the curve. We call this geodesic deviation.

Similarly, two people who begin in love can drift apart without ill intent—simply carried by emotional curvature. Or they could veer into each other's paths again, their geodesics crossing.

Pain often arises from unexpected divergence—the distance between what was and what is.

Maybe understanding your emotional geometry helps you read the paths before pain hits.

ABSTRACT ART: EMOTION MADE VISUAL

In abstract paintings, colour and shape evoke feeling. A splash of red might jolt you; a circle of blue might calm. That's emotion geometry on canvas. No narrative, no explanation—just form that stirs.

Kandinsky believed that shapes and colours had spiritual resonance: a triangle could feel like fire. A circle like eternity. A square-like stability. Art mirrors emotion's shape—and lets us experience it, name it, live it without words.

EMOTIONAL SYMMETRY AND RESONANCE

In symmetry, left and right are mirrored. Emotionally, symmetry can feel like balance—peace, calm, evenness.

Resonance, on the other hand, is when two shapes echo each other. You meet someone and feel an immediate connection—your emotional structures vibrate together. That resonance brings harmony; dissonance brings disquiet.

Emotion's geometry is not only what we feel, but how we share it.

CURVED BEINGS IN EMOTIONAL SPACE

Imagine a three-dimensional grid inside your mind—a space made of memories, senses, projections. Emotions change with motion. Anxiety might pull you forward, grief backward, joy upward.

We are curved beings—curved by our histories, our loves, our traumas. And recognizing our own emotional geometry could help us move differently—choose a different path, reshape the curve, heal the fold.

RESHAPING OURSELVES

If emotions have shape, we can intervene. Mindfulness, therapy, art, music—they are tools to reshape geometry. A breath stretches the curve. A song smooths the plane. A kind act fills a hollow.

That's how change happens—not through willpower, but through reshaping feeling-space.

THE POET'S TASK

To describe emotion is to map interior landscapes. And that's art's domain. Writers, painters, composers—they name: here is a gorge of loss, here the warm rise of hope.

If emotions are geometry, then art is cartography—mapping the invisible shapes of feeling.

Your life is a map you draw moment by moment. Each relationship, each breath, each choice shapes your interior geometry.

FINAL REFLECTION

So, ask yourself: What shape is your heart today?

Does it curve in on itself with pain? Does it open wide in gratitude? Does it spiral in confusion, remain flat in exhaustion, skyrocket in joy? And how will you reshape your geometry so that, in the next moment, your shape fits the one you want to live?

Because the universe is art, and emotions are its language. If time and energy have curves, so do we. If gravity moulds the cosmos, so do emotions mould the soul.

Your geometry is your story.

Think about it.

A CONSCIOUS BORDER

There's a point in every mind, quiet and unsettling, where thoughts stop—not because there is nothing more to say, but because there is **too much** to comprehend. A blank wall. A silent edge. A border.

Not one made of bricks or language. But a border of **consciousness**—where even imagination hesitates, and understanding collapses under its own weight.

Have you ever stood at the edge of a dream you couldn't finish? Or read a riddle so vast it made your thoughts feel small? That's where we begin. At the edge of what we *can* know.

Think about it.

THE LIMITS OF PERCEPTION

Our brain—the magnificent conductor of our reality—is not infinite. It is shaped by evolution to help us survive, not to comprehend the origin of existence or the collapse of multiverses. It sees in limited light. It hears within a narrow band. It thinks in loops. It forgets. It simplifies.

And still, we dare to ask: *What is beyond space? What came before time? What lies after death?*

The irony is beautiful: a finite organ trying to map the infinite universe.

So, what do we do when we hit that wall? When is the question too vast, too paradoxical, too layered?

We find a bridge.

A MIND SEEKING MEANING IN THE ABYSS

When ancient minds looked to the sky, they didn't just see stars. They saw symbols, gods, patterns. Their brains couldn't comprehend the vacuum of space or the curvature of time, but they could **imagine**.

That's the first tool consciousness built: **myth**. Then came **story**, **philosophy**, **art**, and **rituals**.

But there was something still missing. A more precise instrument. A tool not of feeling, but of **form**. Something that could step across the border of limitation—and sketch the shadows of what lies beyond.

That something… was **mathematics**.

MATHEMATICS: THE LANGUAGE OF THE UNKNOWABLE

Mathematics was not *invented*—it was *discovered.* A silent truth waiting behind the veil of perception.

We noticed that numbers described the swirl of stars, the spiral of shells, the fall of apples. We began to write equations not just about what we *could* see—but about what we *couldn't.* We modelled particles we had never touched. We painted shapes that defied three-dimensional space.

Suddenly, we weren't just exploring **reality**. We were expanding **consciousness**.

Math became our telescope into the dimensions we could never walk. A symbolic bridge from what the brain could not hold to what the universe could not avoid.

It became our whisper into the cosmic unknown.

DIMENSIONS WE CANNOT TOUCH

String theory proposes eleven dimensions. Quantum physics suggests particles can exist in multiple places at once. Time bends. Space warps. Logic flips.

Yet… our brain is still wired for three dimensions and linear time. We cannot "see" the fourth dimension.

We cannot feel the fabric of space-time.

We cannot truly imagine infinite parallel

universes, even when equations insist, they might exist.

Our consciousness, as evolved as it is, operates within a very **local** dimension of the cosmos. We are a ripple in a pond trying to comprehend the ocean.

To cross this conscious border, we must first accept that we cannot directly **experience** the higher truths of the universe.

But we can **translate** them.

MATH AS A TRANSLATOR OF CONSCIOUSNESS

If the universe is a great symphony, mathematics is the sheet music.

It's how the music is shared when we cannot hear it ourselves.

From Pi to Planck time, from Fibonacci spirals to gravitational tensors, math holds the keys to a language we are still learning to speak. It doesn't replace consciousness. It **extends** it.

Our minds may not grasp infinity—but calculus gives us limits. We may not perceive a black hole—but Einstein's equations describe its behaviour. We may not feel quantum entanglement—but complex probability and linear algebra let us sketch its strange outline.

Math becomes the geometry of the unknowable.

Think about it.

ARE WE NEARING THE EDGE… OR JUST BEGINNING?

We once thought the Earth was flat. Then a sphere. Then we thought we were the centre. Then, a small part of a galaxy. Then one galaxy in trillions.

Each time we stepped beyond a conscious border, we found not a wall—but another sky.

So now we ask: is this edge—the edge of perception, of brain chemistry, of intuitive understanding—truly the limit? Or just another veil, waiting for a new tool?

Could there be forms of consciousness greater than ours? Could the universe itself be conscious—not like a brain, but like a **field**? Could math be the only one translator, and art, music, emotion, and dreams be others?

We don't know. And maybe we can't.

But the fact that we *ask*—the fact that we press against this invisible boundary—means that we are already stretching it.

THE CONSCIOUS BORDER AND THE SACRED UNKNOWN

There's a reason ancient sages, modern physicists, poets and monks all speak in similar tones when they speak of the ultimate reality.

They reach a point, and then they speak in paradox.

"This is and is not."

"The observer becomes the observed."

"The knower and the known dissolve."

"I am that."

Why? Because they stand at the edge of consciousness. And beyond that edge, words fail. Math continues. Music stirs. Silence speaks.

The Vedas called it *Neti Neti*— "Not this, not that."

The Tao Te Ching begins: "The Tao that can be named is not the eternal Tao."

There is a conscious border. And perhaps the most sacred thing we can do is acknowledge it—not as failure, but as awe.

CLOSING REFLECTION

We are not yet capable of understanding the full architecture of the universe—not because it is flawed, but because **we** are still unfolding.

Our brains were shaped for survival, not omniscience.

Our eyes see light, not dark matter.

Our minds think in years, not aeons.

But through the channel of mathematics—and the reverence of imagination—we reach across the void. We touch what we cannot hold. We sketch what we cannot yet understand.

And maybe one day, as consciousness expands—through evolution, through technology, or through some inner awakening—we will step across that border not with questions, but with calm recognition.

Until then, we must live on the edge. Questioning. Wondering.

And sometimes, simply closing our eyes and feeling the invisible curve of the cosmos against the shape of our thoughts.

Think about it.

THE GOD

Since the dawn of awareness, humanity has looked up, looked within, and asked the same question in a thousand forms: **Is there something—someone—beyond all of this?**

When the stars refused to speak and the mountains stood in silence, we carved our answers into stone, whispered them into fire, and called that silence **God**.

But what—or who—is this God?

Is it a being that created all things? Or a word we created to explain all that we cannot understand?

Is God the artist behind the art face of the universe—or is God the art itself? **Think about it.**

SOMETHING FROM NOTHING? OR EVERYTHING FROM SOMETHING?

Let us start with the foundational question of creation.

The universe exists. Fact.

But did it arise from *nothing*? From a singularity—a point without space, time, mass, or meaning?

Physicists say yes. The Big Bang. A quantum fluctuation in a vacuum. A spontaneous birth from an unstable nothingness.

But *nothing*, by its very definition, cannot contain **potential**. For there to be fluctuation, there must be a field. For a field to exist, there must be space. For space, time must unfold.

So, was it truly *nothing*? Or was it a silence so deep our tools cannot detect it—an **everything** so subtle, it appears empty?

If the universe came from *something*—then what is that something? Is it energy? Information? Or consciousness itself?

Or is that "something" ... God?

THE UNKNOWN AND THE DIVINE PLACEHOLDER

For thousands of years, when humanity reached the limits of its understanding, we often inserted **God** into the unknown.

We did not know what thunder was—so it was the voice of Zeus and Indra. We did not know how the stars moved—so they were the jewels of Vishnu's crown. We did not understand death—so it became a passage to judgment.

God, for many, has been the **gap-filler**. The sacred placeholder for all that lies beyond comprehension.

But as science advanced, the gaps shrank. We now understand thunder. We understand orbits. We can map the brain's electrical response to love and grief.

And so, many ask—has God vanished with the gaps?

Or has God merely moved?

A HIGHER DIMENSIONAL INTELLIGENCE?

Let us consider a possibility—not of myth, but of **dimensionality**.

Imagine a two-dimensional world. Flatland. All that exists is width and height. If a three-dimensional being were to reach into Flatland—say, by placing a finger through the surface—the Flatlanders would see a mysterious circle appear out of nowhere. They'd feel a disturbance. A shape with no explanation. A force beyond.

Now imagine we are the Flatlanders—and something from a higher dimension touches our world.

We feel it as intuition. Coincidence. Miracles. Inspiration. Awe.

Could God be such an entity? Not bound by space or time—not a person with a beard, but an **intelligence** woven into the very structure of reality?

Perhaps what we call "God" is a higher-order consciousness—a **field** we cannot yet decode. Like electromagnetism was before Maxwell. Like gravity before Newton.

And maybe, just maybe, it is not separate from the universe, but the **conscious form of the universe itself**.

Think about it.

PANTHEISM AND THE LIVING COSMOS

The Vedas say, "*Sarvam Khalvidam Brahma*"—all this is Brahman.

Not that God created the universe—but that **God *is* the universe**.

In this view, every particle, every atom, every galaxy is not an object, but an expression of **divine**

vibration. The universe is not a creation, but a manifestation. Not built by God, but *as* God.

Pantheism is not about worshipping trees or stars—but recognising that the same consciousness that breathes in you breathes in them.

That the spark in your eyes and the light of distant suns are reflections of the same source.

This makes God not a judge above—but a pulse within. Not a name—but a **pattern**. Not a being—but **being itself**.

THE SCIENTIFIC DIVINE

Modern physicists sometimes sound like mystics. They speak of symmetry, elegance, the "music of the spheres," the fine-tuned constants that allow life to exist.

Mathematics, too, begins to look like a **sacred language**—not invented, but discovered. Existing independently. Eternal. Like an invisible scripture.

Some scientists believe that if there is a God, it is a **mathematical God**—one that expresses itself in equations, ratios, and fractals. Perhaps the mind of God *is* math.

But even here, a mystery remains. Who, or what, set these constants? Why is the universe not chaos, but balance? Where did the blueprint come from?

Even at its peak, science finds itself looking out into a darkness it cannot illuminate. And there, again, God returns. Not as a defeat of reason—but as a horizon beyond it.

CONSCIOUSNESS: THE DIVINE MIRROR?

Another thread to pull: *What if God is not separate from us at all—but the deepest part of us?*

Carl Jung spoke of the **collective unconscious**—a vast, shared depth beneath the personal self, filled with archetypes, symbols, and ancient instincts.

The yogis speak of **Atman**—the soul—as not separate from **Brahman**, the Absolute.

Mystics in every tradition whisper the same thing: *You are not just a part of the universe. You are the universe, knowing itself.*

Could God be the final layer of our own consciousness—the part we cannot access, yet feel in dreams, visions, meditations?

Could prayer be not begging to the sky—but listening to the **echo of your own cosmic source**?

Think about it.

DOES GOD NEED TO EXIST?

Some ask: Do we *need* God? Can the universe not exist without one?

Perhaps yes. Perhaps not.

But maybe the better question is: What happens to us **when we believe** in God?

Does that belief offer us a new way to live, to forgive, to wonder?

Does it allow us to see beyond anger, beyond ego, into something vaster?

Even if we do not know what God is—*the question itself transforms us*.

And that, strangely and beautifully, may be the most divine thing of all.

FINAL REFLECTION

If the universe is everything, can it truly come from **nothing**?

Or is it the expression of a vast, timeless, unknowable field we have named **God**?

Is God a being? A force? A metaphor?

Or is God the space between the stars, the math behind the light, the question that births all other questions?

In the end, maybe God is not meant to be answered—but meant to be **lived**. Felt. Honoured in silence. Painted in light. Whispered in equations. Reflected in mirrors.

A concept so vast it dances beyond certainty—but close enough to stir your heart in the quiet moments between thoughts. A conscious border. A sacred rhythm. A presence in absence.

…The God.

Think about it.

EPILOGUE

"And so, we come to the edge of our little journey.

We began not with answers, but with questions—questions that ripple through the spine of the cosmos, carried on the breath of stars and stirred in the quiet pulse of thought.

From ***Creation*** *to* ***Consciousness****, from* ***Dreams*** *to* ***Death****, we did not seek to solve the universe, but to* ***feel it****.*

To touch its edge with an artist's brush, to listen with a poet's silence, to question with a seeker's fire.

We asked if the ***universe*** *could be seen not as a cold machine, but as a work of art—alive, breathing,* ***becoming****.*

We wandered through fantasy and reality, time and thought, geometry and God—not to define them, but to dance among them.

Because perhaps ***meaning*** *is not made by standing still, but by moving through the questions with wonder.*

Perhaps what we call truth is not a single dot on the map—

but a thousand overlapping spirals on the canvas of existence.

And you, dear reader—are a stroke of that painting.

A verse in the melody.

A flicker of the light that is also trying to understand itself.

You are not apart from the universe. ***You are the universe, looking inward.*** *You are not just a mind seeking knowledge.*

You are consciousness, drawing borders around infinity—

then daring to cross them.

And maybe that's all we're here to do: To wonder. To create.

To feel. To ask again.

What is this?

Why does it sing?

And who is listening?

The canvas stretches far beyond the page. The questions go on. But now, you carry the brush.

Think about it."

AUTHOR'S NOTE

Born from wonder. Written with reverence.

This book was never meant to explain the universe. It was meant to *listen* to it.

The Art Face of the Universe began as a question I asked myself, staring at the night sky—*"What if the universe wasn't just physics... but poetry?"* What if, beyond equations and forces, there was an aesthetic layer—a rhythm, a feeling, a presence? Not one to replace science, but to sit beside it. A parallel melody.

This isn't a book of facts. It's a book of thoughts—sometimes wild, sometimes grounded, always honest. I've woven together ancient ideas, modern theories, and my own speculative hypotheses not to convince you, but to **invite you**: to reflect, to wonder, to feel.

Along the way, I've drawn from Vedic philosophy, quantum physics, metaphysical psychology, and personal intuition. I've asked questions I can't answer, and perhaps neither can anyone else. But maybe that's the point.

And yes, I used AI to help me refine structure, balance tone, and make the ideas more accessible—but the soul of this book is entirely human. It's mine. And now, if you're holding this in your hands or heart... It's also yours.

Thank you for walking through these pages with curiosity. Keep questioning. Keep wondering. Keep listening.

And always, ***think about it.***

With stardust and reverence,
Navaneet Jaggannathan"

www.ingramcontent.com/pod-product-compliance
Ingram Content Group UK Ltd.
Pitfield, Milton Keynes, MK11 3LW, UK
UKHW021400190726
13845UKWH00028B/797

9 789334 344554